Life's good when the pieces come together.

Published by The Honey Bee Company LLC
2906 Central Street • Suite 258
Evanston, IL 60201
847.864.8032

ISBN 978-0-9861618-1-0

My name is

____

I'm ready to get to work!

This is the Puzzle family.

This is Paige.
Paige loves science.

Science

is the way we find out how things work
and why things are the way they are.

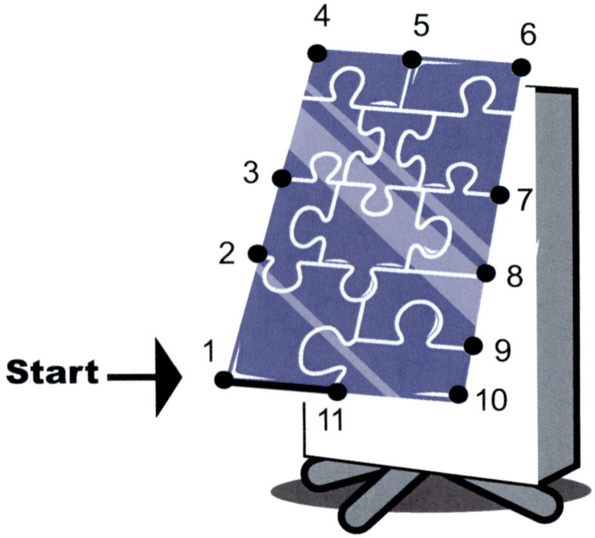

Count the clouds.

I counted _________ clouds.

Scientists

try to find out how things work
and why things are the way they are.

Circle the biggest one.

Paige

tries to find out how things work
and why things are the way they are.

Circle the all the pictures that are the same.

We want to know how a lot of things work.
How many insects do you see?

I counted _________ insects.

Different types of scientists try to find out how different types of things work.

Draw a line to the ones that are the same.

There are scientists who try to find out how things work in this world.

Circle the smallest one.

Help the scientist get to the top of the mountains.
Connect the dots.

There are scientists who try to find out how things work beyond this world.

Circle the biggest one.

Connect the dots.

Scientists love to ask questions.

Which one doesn't belong? Circle the plant that is different.

Paige loves to ask questions.

Scientists love to observe the world.

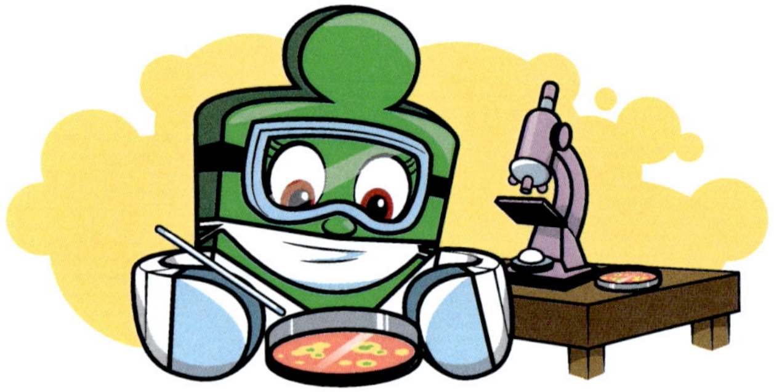

What do you see? Circle the picture that's different.

Paige loves to observe the world.

Count the ants.

I counted __________ ants.

Scientists love to perform experiments.

Paige loves to perform experiments.

Circle the one that is empty.

Paige loves science.
Connect the dots.

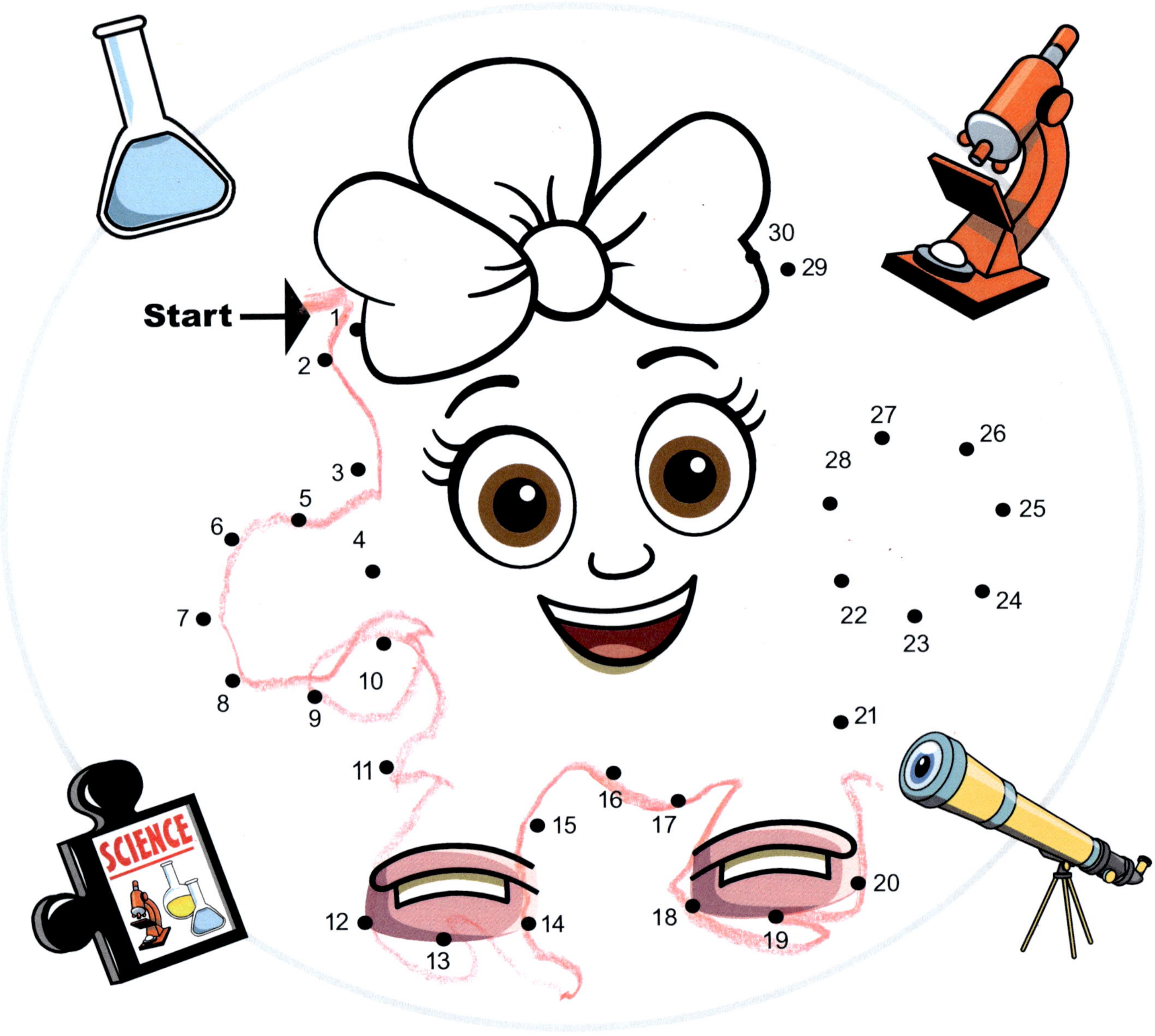

Help Paige get to her science experiment.

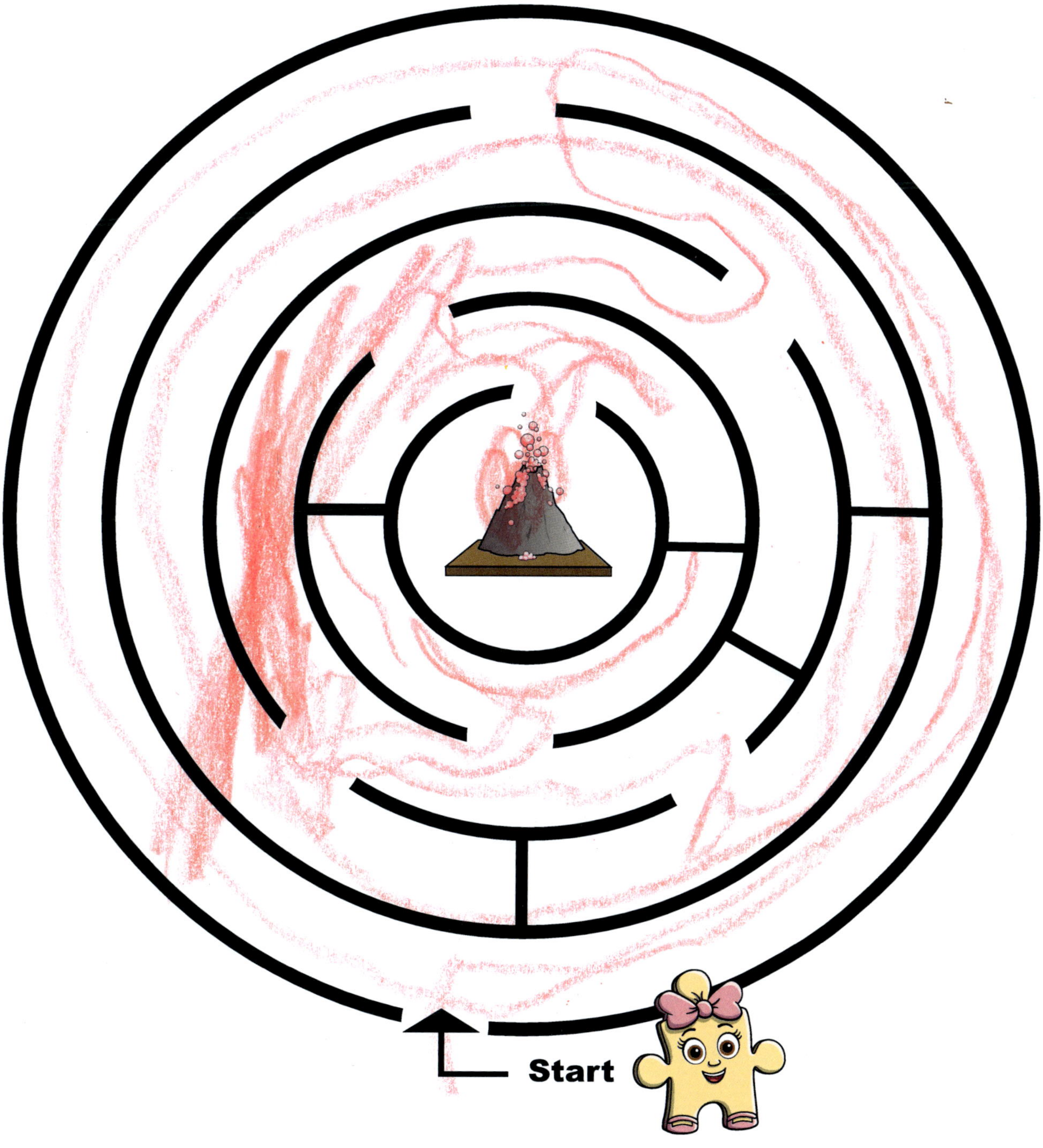

Paige's perfect fit is science.

Draw lines to the pictures that are the same.

What's *your* perfect fit? Draw it here.

Our Mission

Change the course of a nation by making STEM (Science, Technology, Engineering and Math) easy to teach and easy to learn at the very beginning of a child's educational experience

Our Vision

World Domination through STEM

Why We Do What We Do

For Kids: 80% of adults in STEM fields had at least one parent in the field. This means a huge number of our kids are missing out on opportunities that can be life changing.

For Teachers: Classroom teachers face unprecedented challenges to meet expectations and standards. Teachers need time-saving resources that enable them to make small shifts in their classroom lessons to drive STEM interest and create a STEM mindset.

For Our Nation: We're losing the battle. Thirty years ago, 40% of the world's scientists and engineers came from the U.S. Today, it's 15%. STEM fields are where the well paying jobs are—and will be—in the global economy.

Like Us On Facebook

Pin Us On Pinterest

Follow Us On Twitter